DISCOVER!

STRANGE FISH

PUFFER FISH ARE STRANGE

NATALIE HUMPHREY

Enslow PUBLISHING

Please visit our website, www.enslow.com. For a free color catalog of all our high-quality books, call toll free 1-800-398-2504 or fax 1-877-980-4454.

Cataloging-in-Publication Data
Names: Humphrey, Natalie.
Title: Puffer fish are strange / Natalie Humphrey.
Description: New York : Enslow Publishing, 2023. | Series: Strange fish | Includes glossary and index.
Identifiers: ISBN 9781978528420 (pbk.) | ISBN 9781978528444 (library bound) | ISBN 9781978528437 (6pack) | ISBN 9781978528451 (ebook)
Subjects: LCSH: Puffers (Fish)–Juvenile literature.
Classification: LCC QL638.T32 H86 2023 | DDC 597'.64–dc23

Portions of this work were originally authored by Therese M. Shea and published as *Pufferfish*. All new materials this edition is authored by Natalie Humphrey.

Published in 2023 by
Enslow Publishing
29 E. 21st Street
New York, NY 10010

Designer: Tanya Dellaccio
Editor: Natalie Humphrey

Photo credits: Cover Franny Constantina/Shutterstock.com; p. 4 topimages/Shutterstock.com; p. 5 KT photo/Shutterstock.com; p. 7 (red eye puffer) blickwinkel/AlamyImages; p. 7 (eyespot puffer) Pavaphon Supanantananont/Shutterstock.com; p 7 (longnose puffer) Luiz Felipe V. Puntel/Shutterstock.com; p. 7 (mbu puffer) yuliya ivanenko/Shutterstock.com; p. 9 Podolnaya Elena/Shutterstock.com; p. 10 Grisha Bruev/Shutterstock.com; p. 11 Sriploy MacIntosh/Shutterstock.com; p. 13 vkilikov/Shutterstock.com; p. 15 Toxotes Hun-Gabor Horvath/Shutterstock.com; p 17 Franny Constantina/Shutterstock.com; p. 19 KeikoH/Shutterstock.com; p. 21 tarin chiarakul/Shutterstock.com.

Printed in the United States of America

CPSIA compliance information: Batch #CSENS23: For further information contact Enslow Publishing, New York, New York, at 1-800-398-2504.

CONTENTS

Boldface words appear in Words to Know.

PUFFED UP PREDATOR

The puffer fish has one strange way of keeping itself safe. When there's danger, the puffer fish sucks in as much water as it can. This makes this small fish look huge and keeps it from fitting in the mouth of a **predator**! There are more strange things to discover about the puffer fish!

PUFFER FISH ARE ONE OF THE ONLY BONY FISH THAT CAN BLINK!

ONE HUGE FAMILY

There are over 120 **species** of puffer fish around the world. Most live in the salty ocean, but some live in fresh water. Others like brackish water, which is a mix of fresh water and salt water. The only water puffer fish don't like is cold water!

SO MANY PUFFERS!

STRANGE SPIKES

Puffer fish usually have a fat head, small fins, and big eyes. They have rough, tough skin that's often brightly colored with spots or shapes. Some puffer fish have spikes on their skin. These lay flat against the body until a fish puffs up.

THE LARGEST PUFFER FISH CAN GROW TO MORE THAN 2 FEET (60 CM) LONG!

TOUGH TEETH

Puffer fish have four special teeth that never stop growing. These teeth are stuck together and look like a bird's beak. Eating hard things wears their teeth down. If their teeth get too long, they won't be able to chew their food!

PUFFER FISH HAVE STRONG **JAWS** THAT CAN CRUNCH THROUGH A SHELL IN UNDER ONE SECOND.

BLOWN AWAY

Puffer fish are strange hunters. When it's time for dinner, they look for anything moving on the ocean floor. Some puffer fish will even blow water at the sand to dig up their food. When they spot something tasty, they grab it in their mouths and crunch down!

PUFFER FISH HAVE GREAT EYESIGHT AND CAN SPOT **PREY** FROM FAR AWAY.

MEAT EATERS

Puffer fish eat a lot! They eat more than once a day and like crunching animals with tough shells. Smaller puffer fish will eat **plankton**. Bigger puffer fish will eat worms, shrimp, lobsters, sear stars, crabs, and clams. Puffer fish can even swallow shells!

SOME PUFFER FISH HAVE AN EXTRA ROW OF TEETH TO CRUNCH DOWN ON PREY.

POWERFUL POISON

Puffer fish are one of the most **poisonous** creatures on Earth! When in danger, poison can come out of the puffer fish's spikes or out of other body parts. The puffer fish makes this poison from **bacteria** in the food they eat.

TIGER SHARKS AND SEA SNAKES DON'T SEEM BOTHERED BY THE POISON. THEY EAT PUFFER FISH!

TINY BABIES

When it's time to **mate**, the male puffer fish pushes the female to shore. She lays her eggs. They float in the water for about a week before they **hatch**. The baby puffer fish keep their eggshells on their body until their tail, fins, and other body parts are finished growing!

EVEN BABY PUFFER FISH ARE POISONOUS!

PERFECT PETS

Even though wild puffer fish are poisonous, puffer fish sold in stores aren't. Since wild puffer fish poison comes from the food they eat, pet store puffer fish have special food. This makes them safer to keep as pets! Pet puffer fish need to be kept in their own tanks with plenty of space.

THE TAKIFUGU PUFFER FISH IS USED TO MAKE A JAPANESE FOOD CALLED FUGU.

WORDS TO KNOW

bacteria: Tiny creatures that can only be seen with a microscope.

hatch: To break open or come out of.

jaws: The bones that hold the teeth and make up the mouth.

mate: To come together to make babies.

plankton: A tiny plant or animal that floats in the ocean.

poisonous: Having poison, which is something that causes illness or death.

predator: An animal that hunts other animals for food.

prey: An animal that is hunted by other animals for food.

species: A group of plants or animals that are all of the same kind

FOR MORE INFORMATION

BOOKS

Alaraj, Evangelene. *Puffer Fish*. New York, NY: AV2, 2021.

Lukidis, Lydia. *Animals That Are Fish*. Novato, CA: High Noon Books, 2020.

WEBSITES

National Geographic Kids
kids.nationalgeographic.com/animals/fish/facts/puffer fish
Watch videos, see pictures, and learn more facts about the puffer fish!

A-Z Animals
a-z-animals.com/animals/puffer-fish/
Dive into facts about where puffer fish live, what they eat, and more!

Publisher's note to educators and parents: Our editors have carefully reviewed these websites to ensure that they are suitable for students. Many websites change frequently, however, and we cannot guarantee that a site's future contents will continue to meet our high standards of quality and educational value. Be advised that students should be closely supervised whenever they access the internet.

INDEX